To my dearest dad, Salvador Acevedo —B.A.

To my family, friends, and mentors for their endless support —A.G.T.

For my dad, who is the most reliable person in the world and who always knows how to fix tiny to huge things —M.L.A.

Room to Read would like to thank Tatcha™ for their generous support of the STEAM-Powered Careers collection.

Copyright 2022 Room to Read. All rights reserved.

Written by Brittany Acevedo
Featured scientist: Dr. Alina Garcia Taormina
Illustrated by Michelle Laurentia Agatha
Edited by Jocelyn Argueta
Photo research by Kris Durán
Series art direction and design by Christy Hale
Series edited by Carol Burrell, Jamie Leigh Real, Jocelyn Argueta, and Deborah Davis
Copyedited by: Debra Deford-Minerva and Danielle Sunshine

ISBN 978-1-63845-065-8

Manufactured in Canada.

10 9 8 7 6 5 4 3 2

Room to Read
465 California Street #1000
San Francisco, California 94104
roomtoread.org

World change starts with educated children.©

STEAM-Powered Careers

NANOTECHNOLOGY

by **Brittany Acevedo**

featured scientist: **Dr. Alina Garcia Taormina**

illustrated by **Michelle Laurentia Agatha**

Room to Read

Contents

Explore Nanotechnology with Mia and Sunny   6

What Is Nanotechnology?   22

Meet the Scientist   24

Learn More about Nanotechnology   30

Word List   34

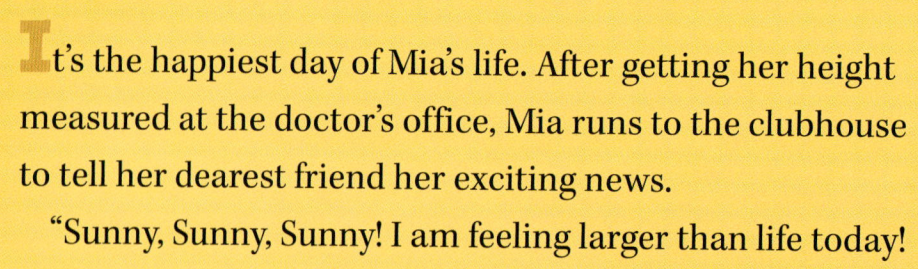

It's the happiest day of Mia's life. After getting her height measured at the doctor's office, Mia runs to the clubhouse to tell her dearest friend her exciting news.

"Sunny, Sunny, Sunny! I am feeling larger than life today! My doctor told me I'm three feet, eight inches tall. Can you believe I'm the tallest in our grade? I am the Magnificent Mia!"

Sunny the snail nervously glides across the clubhouse floor to join Mia by the door.

Cora Mia Jae

Felicia Sunny Bonnie

Nanotechnology

"What's wrong, Sunny?" Mia asks. "You look upset."

Sunny glances down. "Mia, I am so happy that you are growing, but I'm only three inches tall. I'm the smallest student in our grade. I'm the smallest in the world. I am not magnificent. I am Simply Sunny!"

Mia places a hand on Sunny's shell. "Oh, Sunny, you are super, splendid, and spectacular because you are so small. You are not the smallest in the world. Did you know there is a whole field of science that is dedicated to studying the world's smallest objects and materials?"

A measurement has two parts. There is a number and a unit.

STEAM-Powered Careers

Common Units of Measure:
- miles
- feet
- inches
- meters
- kilometers

Sunny's eyes widen. "No way! There is nothing smaller than me."

"You'd be surprised," Mia says. "The field of science is called nanotechnology. Nanotechnology is the study of tiny things that are even smaller than you."

"Whoa!" Sunny says. "How can you measure something smaller than me?"

Mia jumps excitedly and explains, "You can measure small things on the **nanoscale**, like nanotechnologists do."

"What's a nanotechnologist?"

"It's a scientist who studies some of the world's smallest **matter**. It's important to measure small things because everything in the world is made up of small parts that help build larger parts."

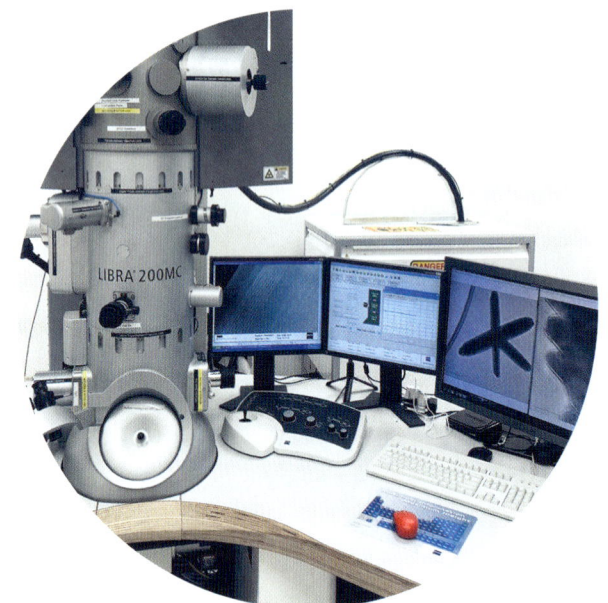

Rulers are a basic tool used for measuring length. Some of the first rulers date back to 2650 BCE

Mia pulls out a ruler and a magnifying glass from her book bag. "Come on, Sunny, the small world awaits!"

Nanotechnology 11

"Slow down, Mia! Where are we going?"

"I'm taking you on a journey to see just how small things in our world can get. There's no better way to see small objects than to shrink down to their size."

Mia and Sunny pause and close their eyes. Moments later, they are both inside a human body.

"Cool!" Sunny shouts. "I'm bigger than a red blood **cell**!"

"Our large body parts are made up of tiny little cells," Mia explains. "Cells are so small that scientists use microscopes to see them. But some things are even smaller than a cell."

Sunny is doubtful. "I don't believe it."

A red blood cell is 7,000 **nanometers** wide. It's so small that you can line up 244 red blood cells across a grain of sand.

Nanotechnology 13

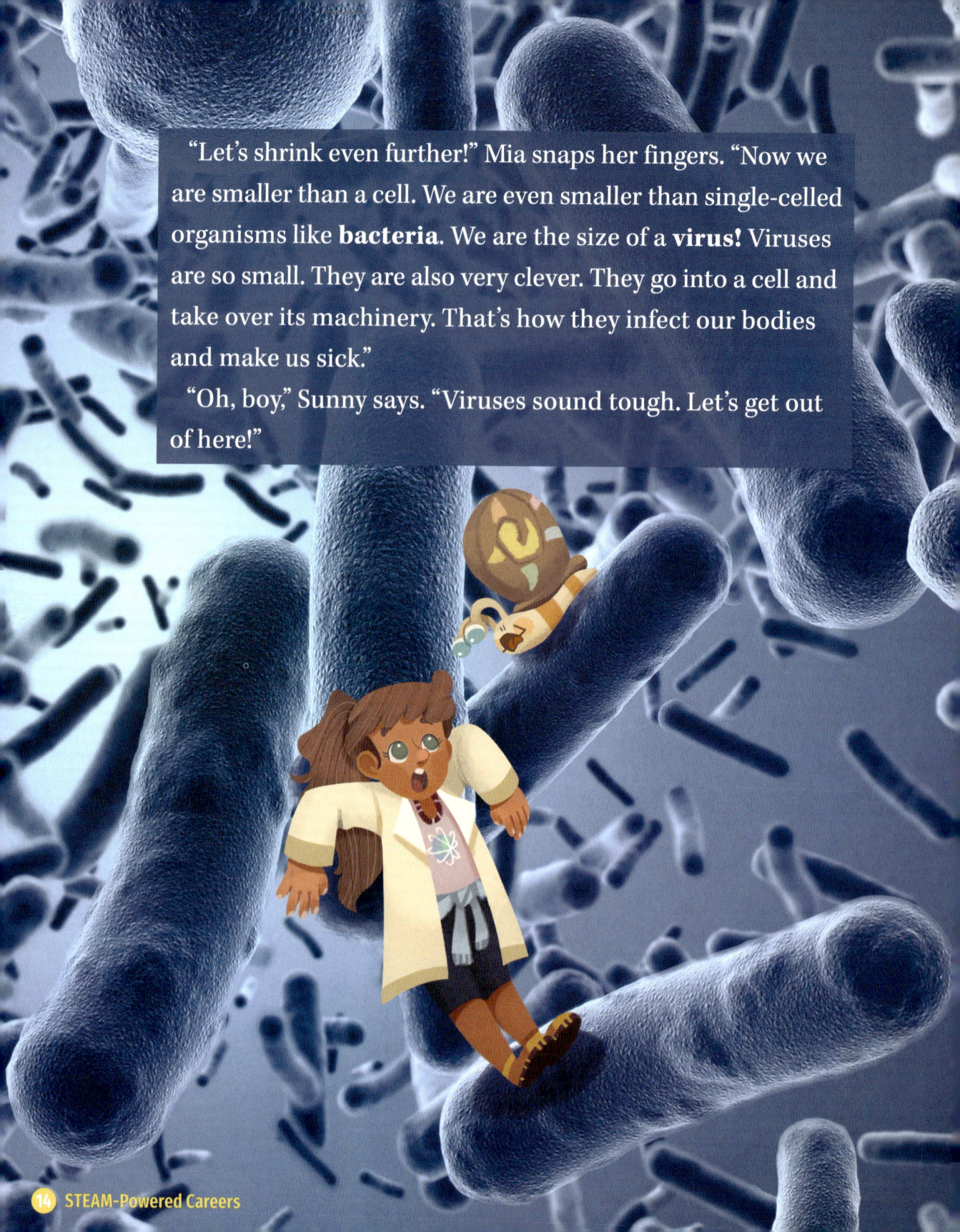

"Let's shrink even further!" Mia snaps her fingers. "Now we are smaller than a cell. We are even smaller than single-celled organisms like **bacteria**. We are the size of a **virus**! Viruses are so small. They are also very clever. They go into a cell and take over its machinery. That's how they infect our bodies and make us sick."

"Oh, boy," Sunny says. "Viruses sound tough. Let's get out of here!"

"We can check out one of the smallest things on Earth: an **atom**!" Mia says. "An atom is the basic unit of matter. Everything is made of atoms. They're so small, you can't even see them with a **light microscope**. That's why we need the **nanoscale**."

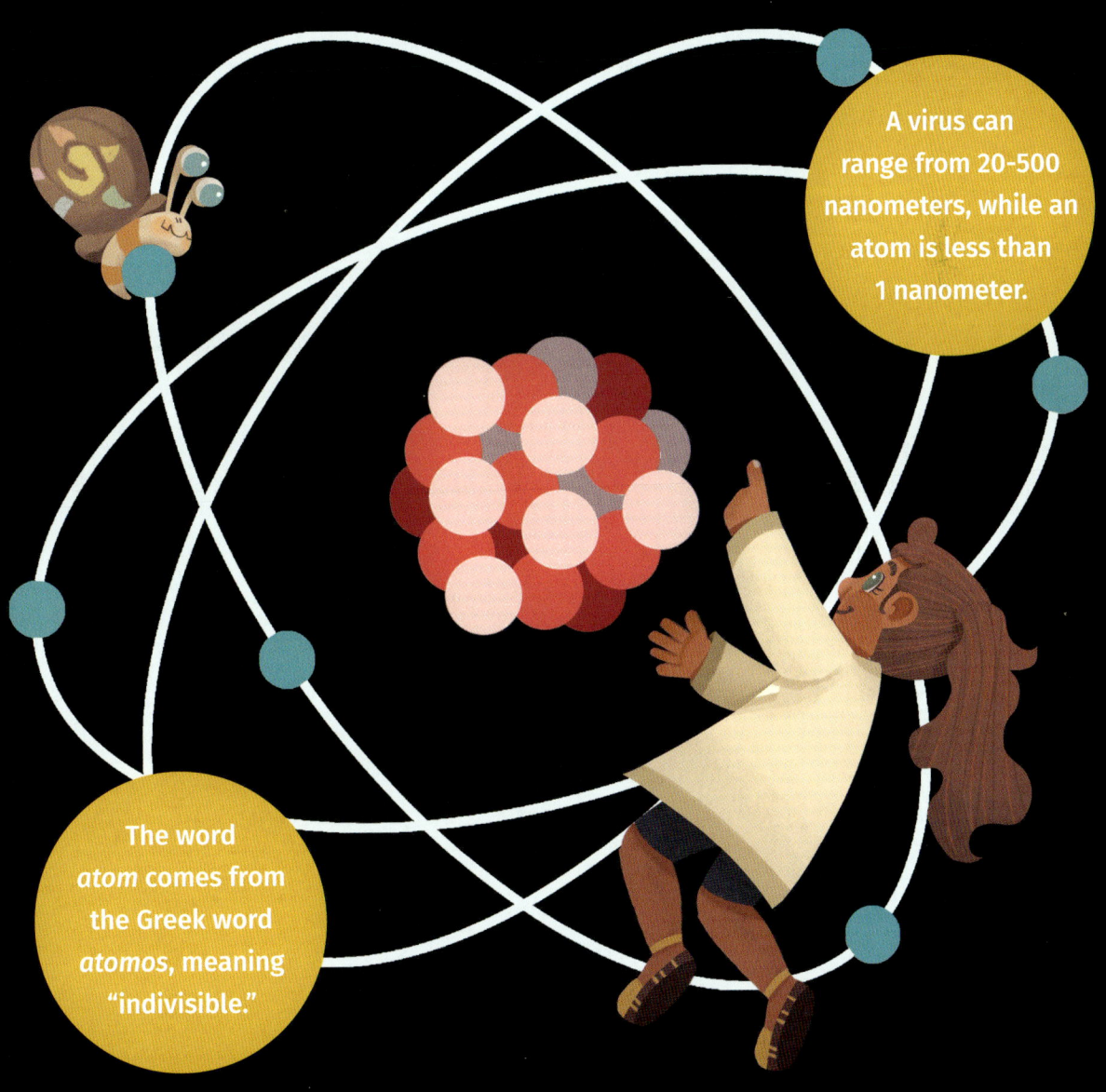

A virus can range from 20-500 nanometers, while an atom is less than 1 nanometer.

The word *atom* comes from the Greek word *atomos*, meaning "indivisible."

Mia and Sunny shrink down again.

"At the nanoscale, we can see atoms," Mia says.

"What's the nanoscale?" Sunny asks.

"Well, the world can be seen using different scales," Mia explains. "On a **macroscale**, objects can be seen with the human eye. Macro items are measured using the imperial system—with inches, feet, and even miles. Other people around the world use the metric system to measure macro items—they use meters, centimeters, and millimeters."

"Wow!" Sunny says.

Ernst Ruska and Max Knoll built the first TEM in 1931.

Water molecule	Glucose	Antibody	Virus	Bacteria
10^{-1}	1	10	10^2	10^3

Nanometers

NANOSCALE   MIC

"On a **microscale**, which is smaller than macro," Mia continues, "people need to use a light microscope in order to see things like cells. At the nanoscale, which is even smaller, scientists need to use a transmission electron microscope (TEM) to see things like atoms. These microscopes are only found in laboratories."

"Cool!" Sunny says. "Tell me more!"

"Let's head to the nano lab!" Mia exclaims, and they zoom out of the human body.

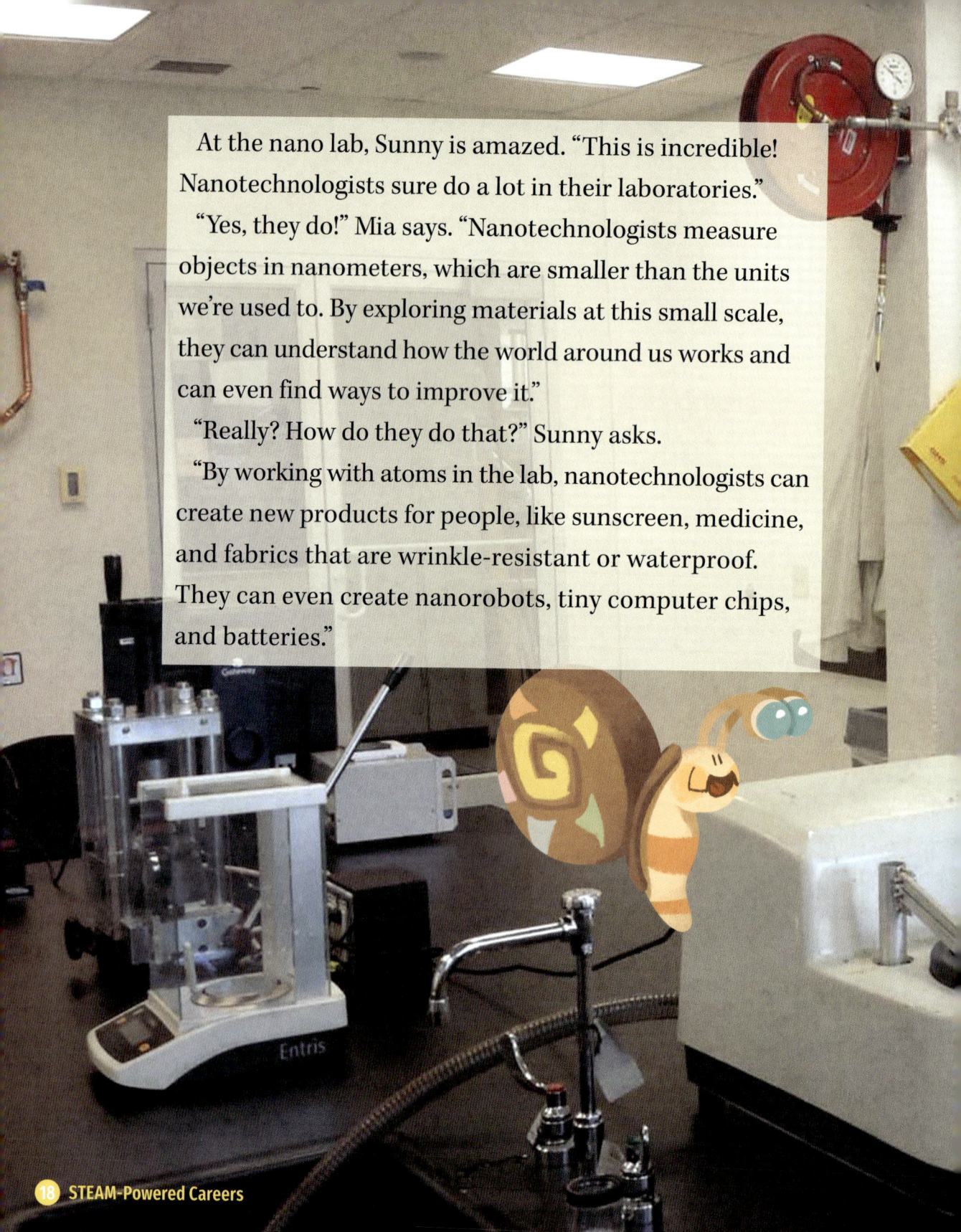

At the nano lab, Sunny is amazed. "This is incredible! Nanotechnologists sure do a lot in their laboratories."

"Yes, they do!" Mia says. "Nanotechnologists measure objects in nanometers, which are smaller than the units we're used to. By exploring materials at this small scale, they can understand how the world around us works and can even find ways to improve it."

"Really? How do they do that?" Sunny asks.

"By working with atoms in the lab, nanotechnologists can create new products for people, like sunscreen, medicine, and fabrics that are wrinkle-resistant or waterproof. They can even create nanorobots, tiny computer chips, and batteries."

Mia and Sunny make it back to the clubhouse just before dinner. Sunny says, "It feels weird returning to such a big world! I guess the smaller things in life can be kind of cool!"

"Kind of?" Mia smiles.

"OK, small measurements are super cool!" Sunny says. "I am ready to rock my small size in the classroom. Thank you so much for showing me the amazing nanoworld. I can't wait for our friends to see how super, splendid, and spectacular I am!"

Mia gives him a big thumbs-up. "You bet!" she says. "Who knows? Maybe one day you'll even be a nanotechnologist!"

What Is Nanotechnology?

Mia and Sunny gave us a macro-level view of **nanotechnology**. Now let's put other parts of the field under our microscopes! Before we tag along with **Dr. Alina Garcia Taormina** for an up-close view of the work she's doing, let's go over some concepts that will be useful in the lab.

Nanotechnology connects science and technology in a unique way.

Nanotechnology — very small / technology: requiring art and skill / the study of

nanotechnologist — person who studies

Nanotechnologists work with materials that are less than one hundred nanometers in size—so the size of a virus and smaller.

Nanotechnologists do many things, including:

- Study matter at the nanoscale level to explore how they can design and create new and improved materials and technology.

- Use different instruments and techniques to build very small materials and devices. It's like building something with the world's tiniest LEGO bricks.

Dr. Alina Garcia Taormina is a materials scientist whose work covers both those areas. Let's ask her some questions, and then she'll show us around the lab.

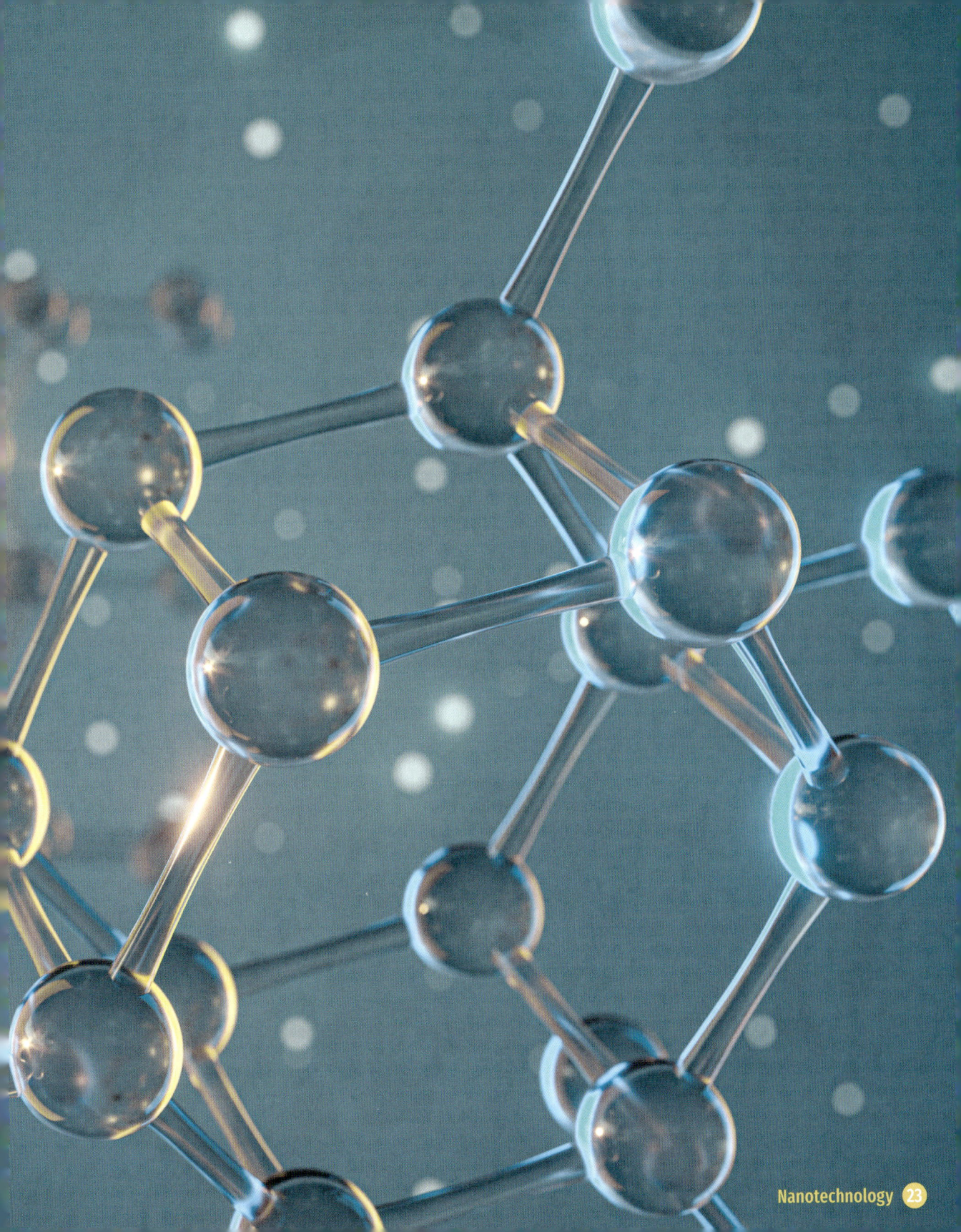

24 STEAM-Powered Careers

Meet the Scientist
Dr. Alina Garcia Taormina

I am a materials scientist who is dedicated to learning about materials on the nanoscale. I have college degrees in chemistry and materials science. I also got a PhD from the University of Southern California (USC), which makes me a doctor of philosophy in materials science.

Fun Fact #1: In college, I studied film production at first, before switching to chemistry! My dream back then was always to be in the film industry.

Fun Fact #2: I love photography! Whenever I feel stressed, I love to pull out my camera and snap a photo.

What is your favorite thing about nanotechnology?

What is your least favorite thing about your field?

Can you show us your daily routine?

There is still so much to learn and discover about the materials that exist at the nanoscale. The nanotechnology field keeps growing!

Even though the nanoworld is small, it can still be a lot to keep up with because there are new discoveries being made every day.

Absolutely! Let's head on over right now!

Nanotechnology 25

A Day in the Nano Lab, Part 1

I start my day in the laboratory with a lot of reading. I am always reading up on what's happening in my research field. My desk is full of papers and books.

I look at unanswered questions in nanotechnology and make a plan to find the answers. Looking for answers is not easy. I have to work together with my peers at USC and with scientists from our partner labs.

26 STEAM-Powered Careers

I usually speak with my lab mates and mentor, Dr. Hodge, to share my thoughts and ask questions. She is my principal investigator, or P.I. for short. I show her my plans for research, and she gives me new ideas to think about. She always guides me in the right direction to find the answers I am looking for.

Research Plan

Problem → Trial
Problem → Solution

A Day in the Nano Lab, Part 2

Once I have a plan for research, I'm ready to start experimenting. I work with nanoscale and microscale 3D-printed materials.

These materials are 3D-printed with features that are down to the nanoscale. That means a real object is printed, but it is so small, we need to use a microscope to see all of its details.

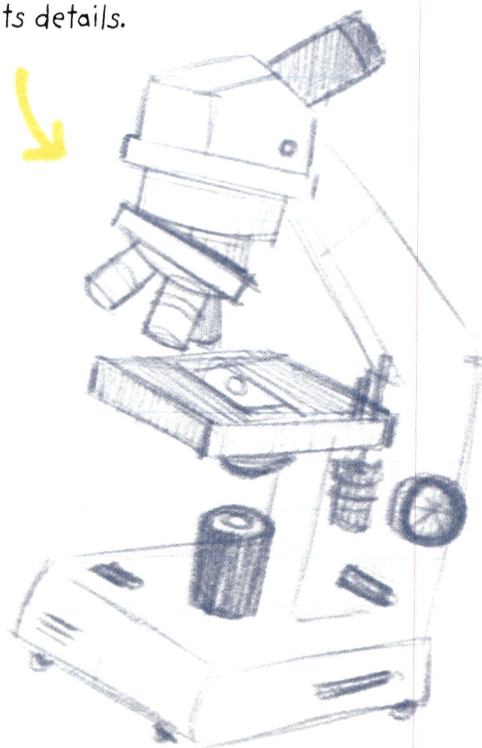

28 STEAM-Powered Careers

I take these nanoscale and microscale 3D-printed materials and coat them using a process called magnetron sputtering. That means I put different metals on these materials. My layers are so small, they are 200 nanometers or less. It's almost like putting layers of frosting on the world's smallest cake.

I use equipment in the lab to closely inspect how good the coating looks on the 3D-printed structures, and I take pictures of them using a scanning electron microscope.

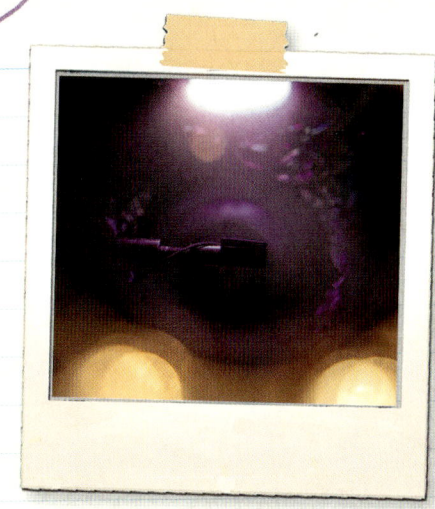

I can see much smaller features with the scanning electron microscope because it uses a beam of electrons instead of a beam of light, which is what the light microscope in a school science lab uses.

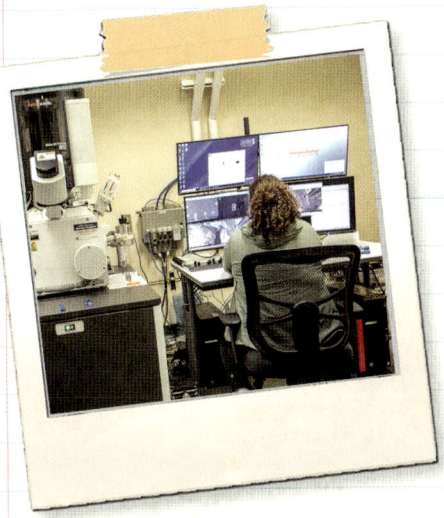

This is my part in the nanotechnology world. I work hard to find answers to big questions.

Nanotechnology 29

Think big when it come to all the jobs involving nanotechnology!

STEAM Careers in Nanotechnology

Understanding how these super-tiny nanomaterials work and can be changed is important because scientists use this information in other fields, like physics, chemistry, biology, medicine science, and engineering.

You may not have heard of nanotechnology because the materials are beyond our ability to see, but this science exists almost everywhere in the world.

Nanotechnologists have built devices known as nanosensors, which collect information about materials on the nanoscale all the way to the macroscale. This device can check the food we eat to see if there is
any risk of finding bacteria, a virus, or a disease in it. This helps keep us safe.

Nanotechnology and engineering are used together to create materials like waterproof fabric, sunscreen, solar panels, and so much more.

The Future of Nanotechnology

Nanotechnology is growing fast. In the next fifteen years, we can expect to see new materials and devices created to help people. Here are a few areas where lots of exciting innovation is happening.

Technology

Computers, machines, and robots will continue to get faster and be able to do more complex things.

Environment

We can also use this nanotechnology to help our planet by cleaning the environment around us, which includes land, air, and water.

Medical

We may also see changes in doctors' offices. For example, nanosensors can be used to go into your body to see if your cells are healthy. This way, scientists can know exactly how much medicine your body needs, and which parts need it the most.

Engineering

We can also expect to see new products made out of nanomaterials, such as more protective sunscreen.

Do You Want to Be a Nanotechnologist?

You can start now!

- Ask a family member, friend, or teacher if they have a magnifying glass or microscope you can borrow. With these tools, look at materials that can't be easily seen with just your eyes—or study familiar things and see what new details you discover!

- Practice putting together different materials to make something new. For example, challenge yourself to make your own slime at home, or even your own robot!

- Ask your family, teachers, and friends about attending science camps and clubs near you—or start your own!

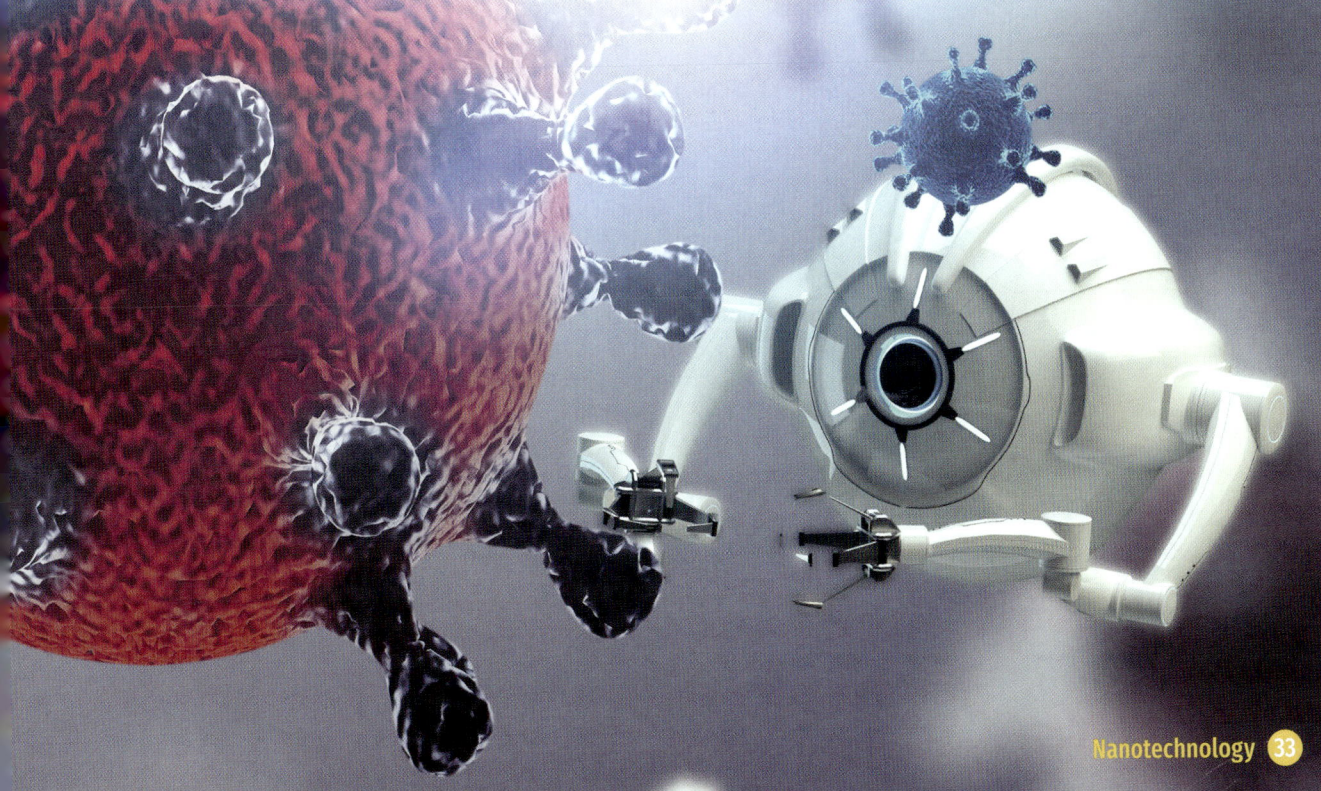

Word List

atom: the smallest unit of *matter* with unique characteristics—every solid, liquid, and gas is made up of atoms in different varieties and combinations

cell: the smallest unit of living *matter*—all living things are made up of cells

light microscope: an instrument that magnifies things up to 1,500 times. A basic unit of *measurement* with light microscopes is the micrometer.

macroscale: a scale used to measure objects that can be seen with the human eye

matter: anything that takes up space

measurement: the size, shape, or amount of something compared to some unit. For example, Mia is 3 feet, 8 inches tall, so the units here are feet and inches.

microscale: a scale used to measure objects that can't be seen with the human eye

nanoscale: a scale of *measurement* that uses nanometers or microns as units of measure

transmission electron microscope (TEM): a special microscope found in labs that magnifies things up to 200,000 times. A basic unit of measurement with TEMs is the nanometer.

virus: a type of germ that causes the flu, colds, measles, and many other diseases

Nanothechnology Resources

Check out this book:

Nano: The Spectacular Science of the Very (Very) Small by Jess Wade

Fun activities and experiments:

http://nanozone.org/how.htm

https://www.nisenet.org/nanodays

https://www.mrsec.psu.edu/content/nano-activities-kids

Acknowledgments

Dr. Alina Garcia Taormina would like to acknowledge her PhD advisor, Professor Andrea Hodge, for her guidance and mentorship throughout her doctoral journey, as well as the National Defense Science and Engineering Graduate (NDSEG) Fellowship Program, for their support.

Brittany Acevedo grew up in Anaheim, California. She is a first-generation Latina American, with a bachelor of arts degree in psychology, a bachelor of arts degree in sociology, and a master's degree in teaching from the University of Southern California. She is a committed public K–8 educator and serves the diverse students and community members of Los Angeles and Orange County as a curriculum developer, author, and teacher.

Dr. Alina Garcia Taormina is from Los Angeles, California, and completed all her schooling there. She was a first-generation college student in her family, earning a bachelor of science degree in chemistry from Loyola Marymount University and a master's in materials science from the University of Southern California (USC). She received her PhD in materials science from the Hodge Materials Research Group at USC. She's excited to start the next journey in her STEAM career as a materials scientist and engineer. As an active member in STEAM-focused outreach and mentoring, she looks forward to continuing these efforts in the next stages of her career.

Michelle Laurentia Agatha was born in Jakarta, Indonesia. Ever since she was young, she has had a huge interest in cartoons and illustrated books. Michelle pursued her dream of becoming an illustrator by earning a Bachelor of Fine Arts degree from the Academy of Art University in San Francisco. Currently, Michelle is working as a children's book illustrator, concept artist, and UI/UX designer.

Explore the Complete

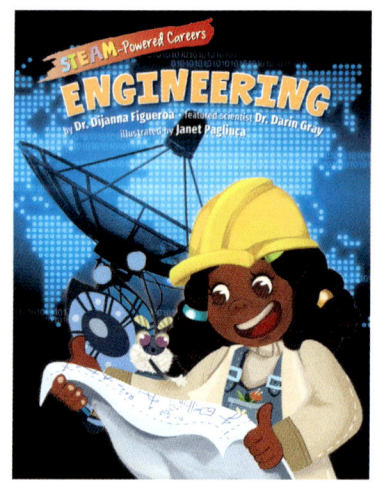

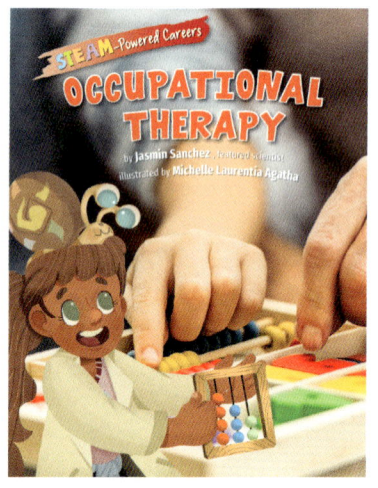
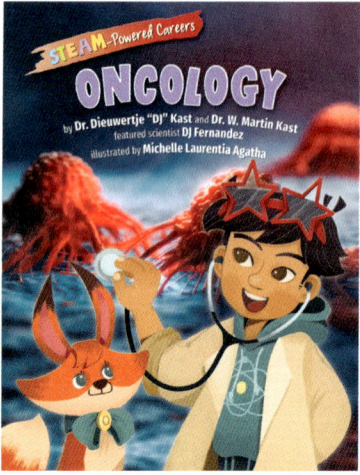

STEAM-Powered Careers Series!

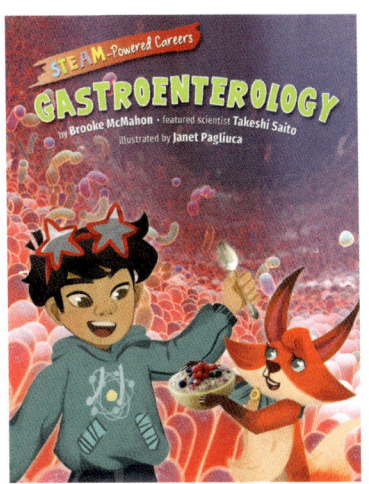

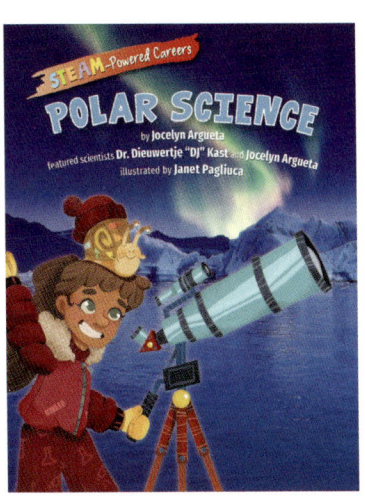
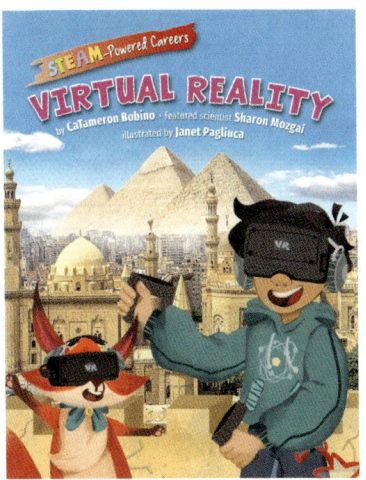

Photo Credits

Cover © Mikhail Rudenko | Dreamstime.com **9** Alexas_Fotos/Pixabay; Gary Greenberg/The Secrets of Sand; Jezper/Depositphotos.com; rukanoga/Depositphotos.com **10** istock.com/the_guitar_mann; NASA Ames/Dominic Hart; WATLab/ University of Waterloo **11** photo by Pixabay from Pexels; istock.com/ChakisAtelier **12–13** Vector8DIY/Pixabay **14** beawolf/Depositphotos.com **18–19** Millersville University **22–23** Vink Fan/Shutterstock.com **24** photos courtesy of Dr. Alina Garcia Taormina and the Hodge Materials Research Group at USC **26–27** Dr. Alina Garcia Taormina and the Hodge Materials Research Group at USC; © Raimond Spekking, CC BY-SA 4.0, via Wikimedia Commons; Agraphe, CC BY-SA 4.0, via Wikimedia Commons; Mersaleashwaran, CC BY-SA 4.0, via Wikimedia Commons **28–29** Dr. Alina Garcia Taormina and the Hodge Materials Research Group at USC; Dr. Alina Garcia Taormina and the Core Center of Excellence in Nano Imaging at USC; Gausanchennai, CC BY-SA 4.0, via Wikimedia Commons; istock.com/HHelene; arz, public domain, via Wikimedia Commons; VIA Gallery, CC BY 2.0 via Wikimedia Commons **30–31** © Pavelbalanenko | Dreamstime.com **32–33** MarkoAliaksandr/Depositphotos.com **34–35** FLPA / Alamy Stock Photo **36–37** © Welcomia | Dreamstime.com; photo courtesy of the Brittany Acevedo; Dr. Alina Garcia Taormina; photo courtesy of Michelle Agatha **40** maykal/Depositphotos.com

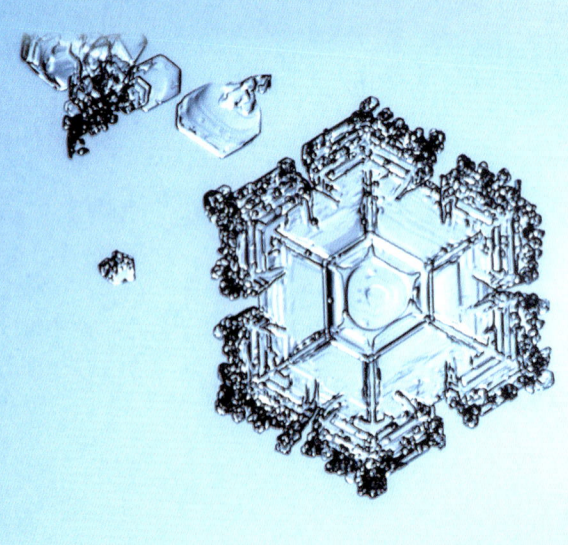

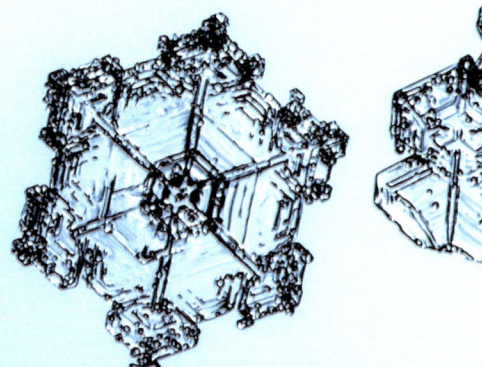